384.55 GIB

Gibson, Diane, 1966-
Television / Diane Gibson.

Date Due	Borrower's Name	
	OCT 23 2001	

384.55 GIB

Gibson, Diane, 1966-
Television / Diane Gibson.

NIPHER LIBRARY

TELEVISION

NIPHER LIBRARY

Published by Smart Apple Media
123 South Broad Street
Mankato, Minnesota 56001

Copyright © 2000 Smart Apple Media.
International copyrights reserved in all countries.
No part of this book may be reproduced in any form
without written permission from the publisher.
Printed in the United States of America.

Photos: pages 7, 8, 9, 11, 12–CORBIS/Bettmann; page
10–CORBIS/Hulton-Deutsch Collection;
page 25–CORBIS/Hulton-Deutsch Collection;
page 30–CORBIS/Roger Ressmeyer

Design and Production: EvansDay Design

Library of Congress Cataloging-in-Publication Data
Gibson, Diane, 1966–
Television / by Diane Gibson
p. cm. – (Making contact)
Includes index.
Summary: Discusses the invention, development,
technology, and future of television.
ISBN 1-887068-64-3
1. Television—Juvenile literature. [1. Television.] I.
Title. II. Series: Making contact (Mankato, Minn.)

TK6640.G53   1999
384.55—dc21   98-41872

First edition

9 8 7 6 5 4 3 2 1

TELEVISION

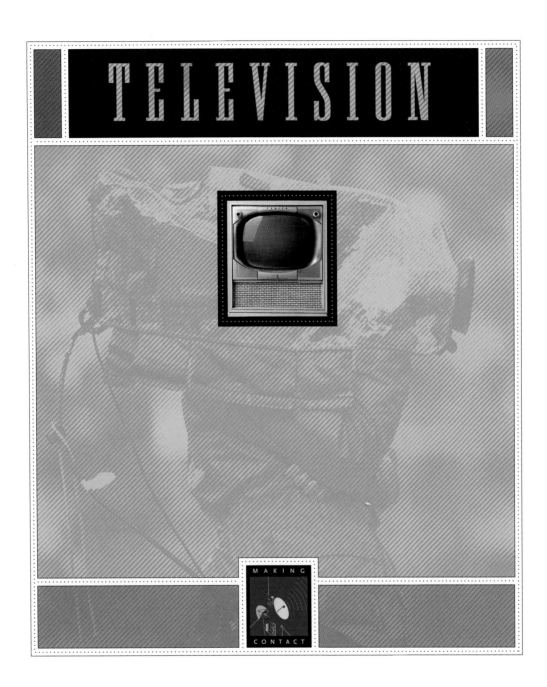

MAKING CONTACT

Diane Gibson

NIPHER LIBRARY

Television is one of the greatest inventions of all time. Although millions of people watch sitcoms, movies, and sporting events on television, it is much more than a source of entertainment. Television brings people news and images of important events from all around the world. This wonderful invention has even taken us beyond our world by allowing people to see detailed pictures of our moon and Mars. By bringing a stunning variety of programs into homes and classrooms, television has become more than a marvel of technology—

it has become a part of our lives.

The Invention of Television

Television, or TV, was not the invention of one person. Today's high-tech television systems are the result of nearly two centuries' worth of scientific discoveries from around the world.

Swedish chemist Jöns Jakob Berzelius made the first important discovery in 1817. He discovered that a chemical element called selenium would carry an electrical current if light was shined on it. The strength of the current depended on the brightness of the light—the brighter the light, the more electricity passed through.

In 1875, an American inventor named G.R. Carey used **photoelectric cells** to build the first television system. He used wires to connect a panel of light bulbs to a panel of photoelectric cells made with selenium. Carey used a lens to focus an image onto the panel of cells. The cells then passed a controlled amount of electricity to each light bulb, resulting in a rough outline of the object shown in the panel of bulbs.

Unfortunately, Carey's invention was complicated, requiring a lot of wires, photoelectric cells, and light bulbs. Even then the pictures weren't very clear. In 1883, inventor Paul Nipkow tried to simplify Carey's invention with an invention of his own: the **scanning disk**.

> Television station WGY in Schenectady, New York, began airing the first scheduled TV broadcasts in May 1928. The programs weren't much fun to watch, however, because they consisted only of moving pictures.

PHOTOELECTRIC CELL
———
a cell whose electrical properties are changed by light

SCANNING DISK
———
a round, flat disk with holes punched into it in a spiral pattern

✻ JÖNS JAKOB BERZELIUS, WHO DISCOVERED HOW TO CREATE ELECTRICAL CURRENTS WITH LIGHT.

> The idea to use orbiting satellites as relay stations for radio, telephone, and television signals was born in the 1940s. The man who came up with the idea was a science-fiction writer named Arthur C. Clarke.

Nipkow began by punching holes into a disk in a spiral pattern. When the disk was spun around, he could see an entire object through the holes. Behind this object was a photoelectric cell connected to one light bulb. Behind the bulb was a second scanning disk, turning at the same speed as the first. Light from the bulb projected an image of the object through the second disk and onto a screen.

Nipkow's system was indeed simpler than Carey's, but the pictures it projected were still far from clear. Clarity was limited be-

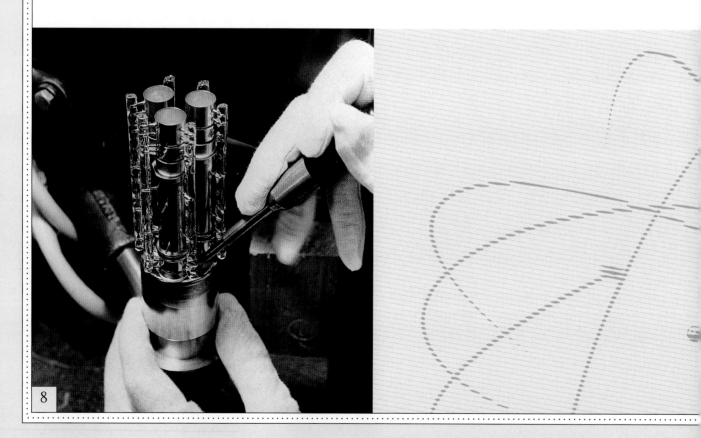

* COLOR TV PICTURE TUBES HAVE THREE ELECTRONIC GUNS TO PROJECT RED, GREEN, AND BLUE COLORS.

cause the disks could only spin at certain speeds, and they could only hold a certain number of punched holes. The next great improvement came in 1897, when German scientist Karl Braun invented the cathode-ray tube. This device used **electrons** to project an image onto a screen. When circling electrons break away from an atom, they create an electric current. Braun discovered that, by pointing a stream of electrons (the cathode ray) onto a screen coated with fluorescent materials, he could make the screen glow. By pointing the stream at the end of a tube, he could make pictures of objects appear on the screen.

By the 1920s, other scientists and inventors were using these discoveries to de-

ELECTRONS

very small particles that circle the center of an atom

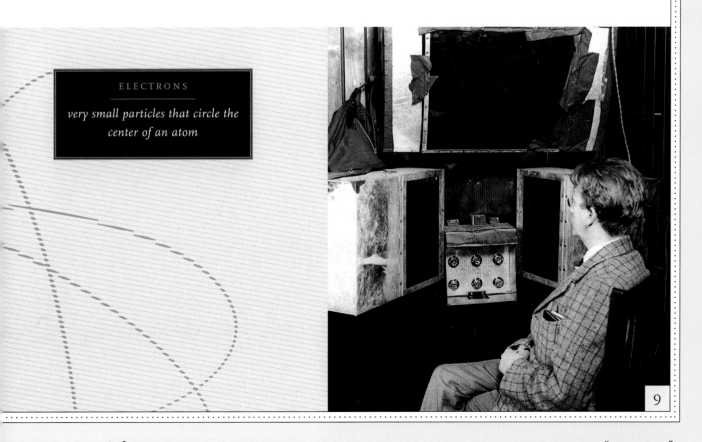

✳ INVENTOR JOHN LOGIE BAIRD, WHO CREATED AN EARLY KIND OF TELEVISION CALLED THE "TELEVISOR."

> The video cassette recorder (VCR) was first developed in 1956 by an American company called the Ampex Corporation. Their VCR was big and bulky, and it used film that was two inches (5 cm) wide. Home VCRs weren't offered to the public until 1963.

velop television systems. The first true television pictures were produced during this decade by John Logie Baird in England and Charles Francis Jenkins in the United States. Their televisions, which were improved versions of Nipkow's scanning disk, used reflected light to create clearer pictures.

Television cameras were also invented and developed during the 1920s. In 1922, American Philo Farnsworth invented an electronic scanning device called an image-dissector tube. This type of camera, which was later used mostly for filming television programs, sent

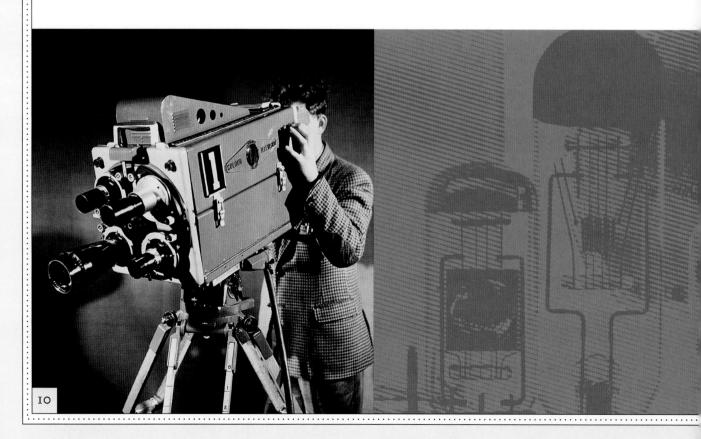

✱ THE ADVENT OF COLOR TV CAMERAS REVOLUTIONIZED THE TELEVISION INDUSTRY.

✶ Philo Farnsworth, who furthered the development of television with his inventions in the 1920s.

images to a light-sensitive screen. It then formed an electronic picture by releasing electrons onto the screen.

Vladimir Zworykin, a Russian-born scientist who later moved to the United States, soon created the ionoscope, a device containing a three-inch by four-inch (7.5 cm by 10 cm) plate that held thousands of tiny photoelectric cells. The cells produced an electric charge whenever Zworykin focused a light image on them. A stream of electrons scanned the cells just as human eyes scan reading material—from left to right and top to bottom. This process released the electronic charges in a certain order. Then the charges were **amplified** and sent to a transmitter. Zworykin's incredible invention was later used for filming movies.

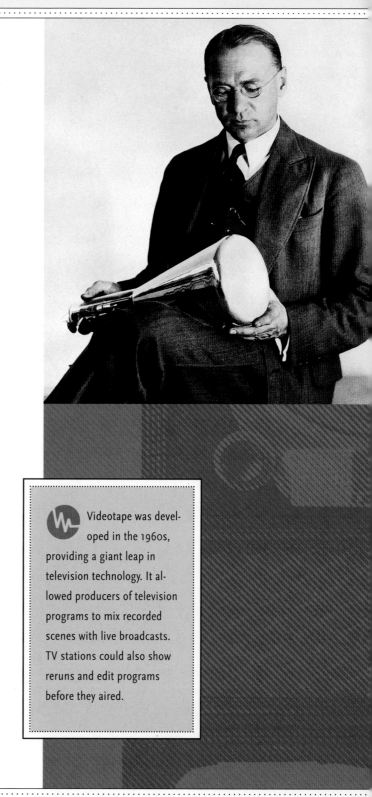

Videotape was developed in the 1960s, providing a giant leap in television technology. It allowed producers of television programs to mix recorded scenes with live broadcasts. TV stations could also show reruns and edit programs before they aired.

✳ VLADIMIR ZWORYKIN HOLDING A CATHODE-RAY TUBE, AN INVENTION THAT HE IMPROVED UPON.

Almost all of the pictures that had been created up to this point were black and white. Although John Logie Baird projected color images during tests he conducted in 1929, the idea didn't really catch on in the United States until the 1950s.

Over the next several decades, new inventors stepped forward to improve these inventions again and again. New discoveries such as communications satellites and video cassette recorders (VCRs) came along, improving television's capabilities and increasing the public demand for televisions. What began in a little lab in Sweden almost 200 years ago has now become an indispensable marvel of technology found in virtually every home in America.

AMPLIFIED

made stronger or louder

It Starts With a Camera

When someone turns on a television, pictures fill the screen almost immediately. But before we can see those pictures, television signals must be created with a camera. Then those signals must be transmitted, or sent, to the television.

During filming, a camera's lens focuses on the image in front of it. To create color signals, mirrors or light filters inside the camera take a full-color scene and separate it into three single-color pictures: red, green, and blue. These are called the primary colors of light. The video signals are produced by **image sensors** inside the camera. If the film is black and white, the camera has only one sensor. For color films, the camera must have at least three sensors—one for each primary color.

One type of image sensor is called a **charge-coupled device (CCD)**. Most home video cameras contain this kind of sensor, which is a silicon chip similar to the chips inside a computer. The sensor includes elements—substances made up of atoms—on its surface that conduct electricity. When light shines from the lens onto the sensor's elements, the light causes electrons to flow into the **capacitor**—the greater the amount of light, the stronger the

> NASA launched *Telstar 1*, the first TV communications satellite, in 1962. Later that year, *Telstar 1* relayed live TV signals across the Atlantic Ocean for the first time.

IMAGE SENSOR
a device that changes light images into video signals

CHARGE-COUPLED DEVICE (CCD)
a silicon chip that changes light into an electrical image

CAPACITOR
a device that collects and stores electrical charges

✳ SILICON CHIPS LIKE THIS ONE ARE OFTEN USED IN CAMERAS TO STORE VIDEO SIGNALS.

electrical charge in the capacitor. When the CCD releases the stored-up charge, a video signal is created.

The CCD releases its charge in a scanning pattern: from left to right and top to bottom. In the United States, television cameras use a scanning pattern with 525 numbered lines. Generally, all the odd-numbered lines are scanned first, then all the even ones. When all 525 lines have been scanned, a television picture—known as a frame—is complete. In order to give us pictures that move

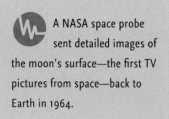

A NASA space probe sent detailed images of the moon's surface—the first TV pictures from space—back to Earth in 1964.

ENCODER

an instrument inside a TV camera that changes red, green, and blue video signals into one unified signal for broadcasting

smoothly, a camera has to scan at least 30 frames in one second. That means the camera must scan 15,750 individual lines or more per second!

After they are scanned, the signals are amplified and sent to an **encoder** inside the camera. The encoder combines all signals into one unified signal so that televisions can decipher the information and produce color pictures. This collection of signals may be recorded on tape for viewing at a later date or may be sent straight to our homes for a live broadcast.

* MODERN VIDEO CAMERAS ARE VERY SOPHISTICATED YET COMPACT AND EASY-TO-USE.

Broadcasting Signals

Once a camera has created television signals, they are ready for broadcasting. Some TV stations send the signals directly through the air. The signals go through an instrument called a transmitter, which separates the audio (sound) and video (images) signals and converts them into radio waves. These waves travel at the speed of light: about 186,282 miles (300,000 km) per second. However, they can travel only in straight lines. Because the earth is round, radio waves head off into space after about 150 miles (250 km). To send them farther, other means—such as cables, satellites, or **relay stations**—must be used.

> TV masts, or antennas, are sometimes used to send and receive television signals in the air. The tallest mast belongs to KTHT-TV in Fargo, North Dakota. It reaches 2,063 feet (629 m) into the sky.

Television stations use different bands, or **channels**, to broadcast their programs. In the United States and Canada, there are 68 channels available for use. These channels are divided into **Very High Frequency (VHF)** and **Ultrahigh Frequency (UHF)** channels. VHF stations use channels 2 through 13, and UHF stations use channels 14 through 69.

Instead of sending TV signals through the air, some stations send their signals through **coaxial cable**. Cable transmission is especially useful for bringing clear television pictures to people who can't put

an antenna on their roof. Many people in large cities subscribe to cable TV. By paying a monthly fee, they receive a range of channels that bring them sports, movies, music, and more. This coaxial cable system has also been used for many years in telephone lines.

Many television companies are now switching from coaxial cable to **fiber-optic cable** for improved picture quality. Fiber-optic cables allow faster transmission of more information. In fact, they can carry 30,000 times more information than a regular telephone line.

RELAY STATION

a tower that receives and passes on broadcast signals

CHANNEL

the band of radio-wave frequencies a television station uses to broadcast its signals

VERY HIGH FREQUENCY (VHF)

a range of radio-wave frequencies used by television channels 2 through 13

ULTRAHIGH FREQUENCY (UHF)

a range of radio-wave frequencies used by television channels 14 through 69

✳ THE TINY BUT POWERFUL GLASS TUBES OF FIBER-OPTIC CABLES ARE REPLACING MANY TRADITIONAL TV CABLES.

Sometimes television signals are converted into microwaves, which are a type of radio wave. Microwaves are relayed from tower to tower on their way to viewers' homes. These towers, which are usually spaced about 30 miles (48 km) apart, automatically receive, strengthen, and pass along the signals.

To transmit television signals across great distances, such as across oceans, communications satellites are used. These satellites usually fly in a **geostationary orbit** around the earth at about 22,000 miles (35,700 km) above the equator. They can relay a TV signal halfway around the world almost instantly. If three satellites are used to form a huge communications triangle, a picture can be sent anywhere in the world in less than half a second. This amazing technology was demonstrated

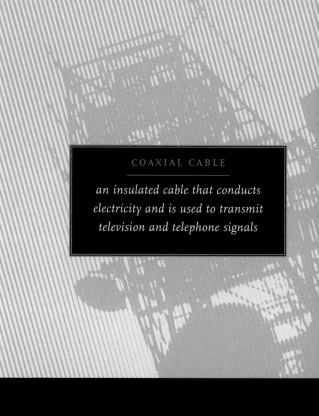

COAXIAL CABLE

an insulated cable that conducts electricity and is used to transmit television and telephone signals

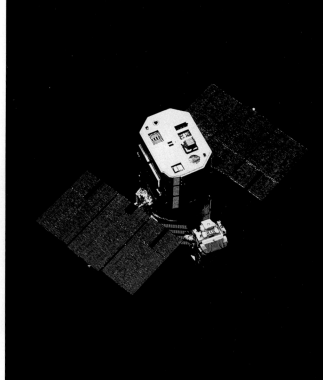

* HUNDREDS OF COMMUNICATIONS SATELLITES IN SPACE TODAY RELAY TV SIGNALS AROUND THE WORLD.

during the opening ceremonies of the 1998 Winter Olympics in Japan, when satellite-assisted television allowed choirs from countries around the globe to sing together in unison.

People who subscribe to satellite television get their programs directly from satellites in space. Subscribers receive TV signals with dish-shaped antennas mounted on their homes or positioned in their yards. Not long ago, these dishes were about 10 feet (3 m) across. But as satellite transmissions became more powerful, the dishes became smaller. Today, home satellite TV dishes are 18 inches (48 cm) or less in diameter. For a monthly fee, satellite TV viewers can choose from hundreds of channels, watching programs sent from stations around the world.

> **FIBER-OPTIC CABLE**
>
> *a cable that contains thin, clear fibers of glass or plastic, through which TV and telephone signals are transmitted*
>
> **GEOSTATIONARY ORBIT**
>
> *the orbit of a satellite above the equator; because the satellite moves at the same speed as the earth, it appears to stay in one place*

✷ DISH ANTENNAS RECEIVE TV SIGNALS FROM SATELLITES THOUSANDS OF MILES FROM EARTH.

How a Television Works

To pick up television signals broadcasted through the air, people need **aerials** on or near their roofs. Aerials—which can be a satellite dish or a standard antenna—pick up the signals and pass them along to televisions.

Signals are passed from the aerial into a television's **tuner**. The tuner is receptive only to signals from the station a viewer is watching. The viewer decides which station to "tune in" by selecting a channel using the remote control or buttons on the TV. Some televisions have two tuners: one for VHF channels and one for UHF.

Once the proper signals are tuned in, electronic circuits amplify them and separate audio signals from video signals. The audio waves go to the speakers, where they are changed into the sound waves we hear. The video waves go to a **decoder**, which changes the video signals into the three primary colors—red, green, and blue.

Next, the decoder sends the video signals to the **picture tube**. The outside of the picture tube is the glass screen that we see. The inside

> PBS aired the first national TV program with captions on August 15, 1972. Today, many nationally broadcast TV programs feature captions. These captions, which are written versions of the words being spoken on-screen, allow hearing-impaired people to enjoy the programs.

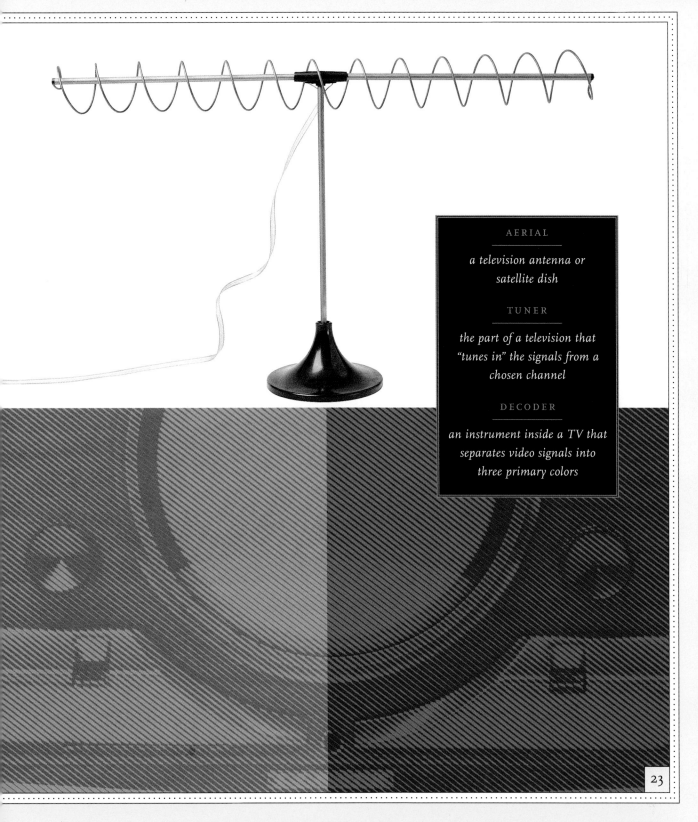

AERIAL

a television antenna or satellite dish

TUNER

the part of a television that "tunes in" the signals from a chosen channel

DECODER

an instrument inside a TV that separates video signals into three primary colors

✲ A TRADITIONAL TV ANTENNA—THIS TYPE OF AERIAL HAS LARGELY BEEN REPLACED BY DISH ANTENNAS AND CABLES.

> VCRs didn't become popular until the 1970s, when the J. Victor Company of Japan developed thinner, easier-to-use VHS tape. This tape was just half an inch (12.7 mm) wide. VCR sales skyrocketed during the 1980s, when the machines also became smaller and easier to use.

of the screen is coated with a thin layer of colored **phosphors**. These phosphors are in the form of tiny dots. An average-sized color television screen contains more than 600,000 of these dots. They are arranged in vertical rows that alternate in color: red, green, blue, red, green, blue, and so on.

The other end of the picture tube has three electronic guns—one for each primary color. Each gun shoots a stream of electrons at the screen. The energy from the electrons causes the phosphors to glow, creating a picture. To do this, the guns must scan the

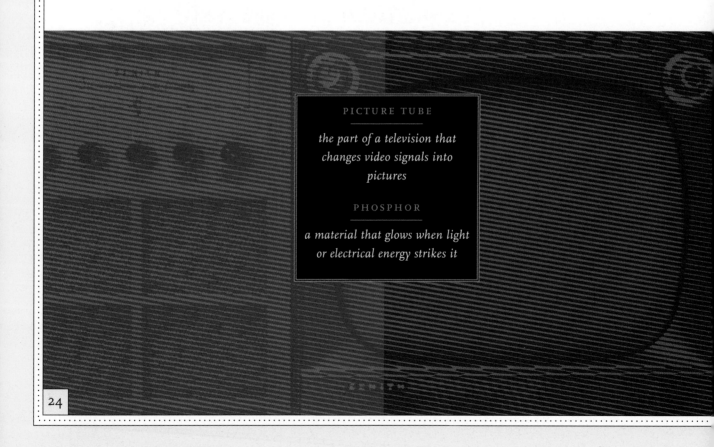

PICTURE TUBE

the part of a television that changes video signals into pictures

PHOSPHOR

a material that glows when light or electrical energy strikes it

screen in the exact sequence the camera used when the scene or action was originally filmed.

Each gun shoots electrons at only one color of phosphors. The red gun shoots at red phosphors, the green gun shoots at green, and the blue gun shoots at blue. To eliminate misfires, a metal plate called a **shadow mask** is positioned behind the screen. The guns must shoot through this plate to scan the phosphors. The shadow mask contains thousands of tiny holes that keep the electron beams from hitting the wrong colors.

> By 1988, the four major American TV networks (ABC, CBS, NBC, and FOX) were broadcasting their programs in stereo. Most TV and VCR manufacturers were also building stereo technology into their products by that time.

* SINCE AS EARLY AS THE MID-1950S, TELEVISION MANUFACTURING HAS BEEN A HUGE BUSINESS.

The strength of the electron beam determines how brightly the phosphors glow, with a stronger beam producing brighter color. Looking at a TV screen through a magnifying glass would reveal tiny dots of only red, green, and blue. However, viewers watching the screen from a normal distance see many colors. As those tiny colored dots blend together, our eyes cannot keep them separate. This mixing of primary colors registers in our brains as a wide range of colors.

SHADOW MASK

a metal plate punched with a pattern of holes that keeps electron beams from hitting phosphors of the wrong color on a TV screen

✶ Magnified images of a television screen show the shadow mask that separates electron beams.

A New Millennium

The last two decades of the 20th century have introduced many improvements in television technology. Researchers have continued to develop televisions with clearer, sharper pictures. One approach has involved combining two modern technologies—computers and television. The result is digital TV.

A digital television receives signals and changes them into **binary code**, which is the language used by computers. A digital TV offers viewers crisp sound and an amazingly clear picture. Digital systems also allow TV signals to carry much more information. This new technology is paving the way for the televisions of the future. In fact, by the late 1990s, President Bill Clinton had encouraged all television stations in the United States to be broadcasting digital signals by the year 2006.

The key to high-quality pictures is the number of lines scanned by the video camera and television—the more lines scanned, the better the picture. Systems that use the old 525-line scanning patterns are quickly being replaced by **high-definition television (HDTV)**. HDTV

> Digital televisions went on sale to the public in the fall of 1998. This type of television can receive digital signals, which carry 33 percent more information than non-digital signals.

systems scan more than 1,000 lines for each frame, offering crystal-clear pictures.

An even better system is **improved-definition television (IDTV)**. IDTV is very similar to HDTV, but there is one important difference. HDTV and older systems create one frame of a television picture by scanning all the odd-numbered lines first, then all the even-numbered lines. An IDTV system scans all the lines at the same time, producing pictures that are even more lifelike than those produced by HDTV.

> **BINARY CODE**
> *a computer language based on combinations of the numbers 1 and 0*
>
> **HIGH-DEFINITION TELEVISION (HDTV)**
> *a television system that uses more than 1,000 scanning lines to produce sharp pictures*

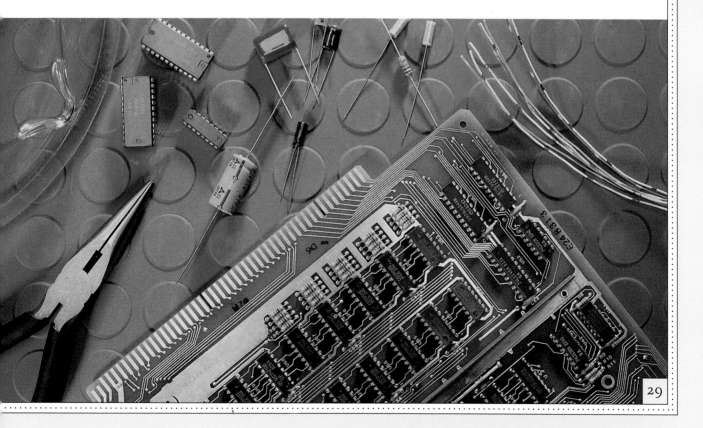

✳ TINY ELECTRICAL CIRCUITS HELP GIVE TODAY'S TELEVISIONS SHARPER PICTURES THAN EVER BEFORE.

In the ongoing quest to make TV screens clearer and lighter, researchers are perfecting the liquid crystal display (LCD) screen. This screen, which uses very little power, is similar to the display screen used in digital watches. These displays are filled with liquid crystals that either block light or let it pass through, depending upon the electrical signals they receive. A screen with LCD capability is very thin and can even hang on a wall like a picture. With an LCD screen and modern surround-sound speakers, the day may soon come when your living room will feel like a movie theater.

Some companies are also developing compact televisions for personal use and

✳ High-definition television developer Yves Faroudja displaying a TV's circuit board.

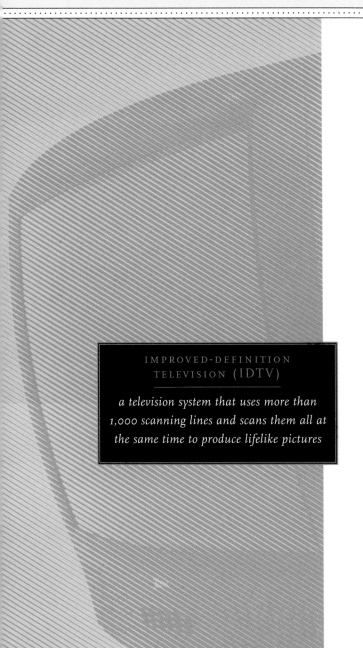

IMPROVED-DEFINITION TELEVISION (IDTV)

a television system that uses more than 1,000 scanning lines and scans them all at the same time to produce lifelike pictures

travel. These TVs may be as small as 14 inches (35.5 cm) wide and have screens that are flatter than those of a laptop computer. There are also television systems available that allow people to access the Internet through their TV. Some day, home computers, televisions, and stereo systems may all be combined into one system.

Television has come a long way from the old system of wires, light bulbs, and selenium cells that delivered a fuzzy image in 1875. It has become an advanced technology that we rely on every day for news, information, and entertainment. The design and uses of television today promise to be just the beginning of this technology's limitless future.

INDEX

A
Ampex Corporation 10

B
Baird, John Logie 9, 10, 13
Berzelius, Jöns Jakob 6, 7
binary code 28, 29
Braun, Karl 9

C
Carey, G.R. 6, 8
cathode ray 9, 12
charge-coupled device (CCD) 14, 15
Clarke, Arthur C. 8
Clinton, Bill 28
communications satellites 8, 13, 14, 18, 20–21

F
Farnsworth, Philo 10, 11

I
image-dissector tube 10
ionoscope 12

J
Jenkins, Charles Francis 10

L
liquid crystal display (LCD) screens 30

M
microwaves 20

N
NASA 16
Nipkow, Paul 6–8, 10–12

P
photoelectric cells 6, 7
Public Broadcasting System (PBS) 22

R
radio waves 18

S
scanning disc 6–8

T
television
 aerials 22, 23
 cable 18–19, 20, 21
  coaxial 18–19, 20
  fiber-optic 19, 21
 cameras 10–12, 14–17
 color 24–26
 compact 30–31
 design (equipment) 22–26
 digital 28
 high-definition (HDTV) 28–29
 improved-definition (IDTV) 29, 31
 masts (antennas) 18
 networks 25
 satellite 21
  dishes 21, 22
 signals 14, 18
Telstar 1 14

U
Ultrahigh Frequency (UHF) 18, 19, 22

V
Very High Frequency (VHF) 18, 19, 22
video cameras 14, 17
video cassette recorders (VCRs) 10, 13, 24
videotape 12

W
Winter Olympics (1998) 21

Z
Zworykin, Vladimir 12